HOW TO COUNT TO INFINITY

Marcus du Sautoy is Professor of Mathematics at the University of Oxford where he holds the prestigious Simonyi Chair for the Public Understanding of Science and is a Fellow of New College.

Du Sautoy has received a number of awards for his work including the London Mathematical Society's Berwick Prize for outstanding mathematical research and the Royal Society of London's Michael Faraday Prize for 'excellence in communicating science'. He has been awarded an OBE for his services to science and was recently elected a Fellow of the Royal Society.

His mathematical research has covered a great many areas including group theory, number theory and model theory, but he has been equally successful in his promotion of mathematics to the general public. He has published a number of best-selling, non-academic books and appears regularly on television and radio.

www.simonyi.ox.ac.uk

Other titles in the series:

How to Play the Piano James Rhodes

How to Draw Anything Scriberia

How to Land a Plane Mark Vanhoenacker

How to Understand E=mc² Christophe Galfard

HOW TO
COUNT TO INFINITY

MARCUS DU SAUTOY

Quercus

First published in Great Britain in 2017 by

Quercus Editions Ltd
Carmelite House
50 Victoria Embankment
London EC4Y 0DZ

An Hachette UK company

A CIP catalogue record for this book is available
from the British Library

ISBN 978 1 78648 497 0

Jacket design by Setanta, www.setanta.es
Jacket illustration by David de las Heras
Illustrations by Amber Anderson
Author photo © Oxford University Images / Joby Sessions

10 9 8 7 6 5 4 3 2 1

Text designed and typeset by CC Book Production
Printed and bound in Great Britain by Clays Ltd, St Ives plc

Woody: Hey, Buzz! You're flying!

Buzz: This isn't flying, this is falling with style!

Woody: To infinity and beyond!

Toy Story

Contents

Introduction: Ready, Steady, Go   1

1   How Long Have We Been Counting?   5

2   Counting Faster and Faster   13

3   Welcome to the Infinity Hotel   21

4   Irrational Numbers   29

5   My Infinity Is Bigger than Your Infinity   43

Ready, Steady, Go

How to count to infinity? It couldn't be simpler. You start at 1 and then keep going. 1, 2, 3, 4 . . . The only trouble is it's going to take you quite a long time . . . especially towards the end (to steal a Woody Allen joke). In fact, you're never going to get there. Time will run out. The Polish artist, Roman Opalka, tried to paint all the numbers from 1 to infinity. He started in 1965. He got as far as 5,607,249, and then died in 2011 before he painted the next number.

Even if you said the numbers out loud rather than painting them, the likelihood is that in a lifetime you might make it as far as one billion and then you'll gasp your last breath and fail to make it to one billion and one. And that's provided no one interrupts you. Lose your place and it's back to square (or number) one. But even if you only make it to one billion, you'll know there's always another number waiting out there for anyone who can make it a bit further. A trillion. A zillion. A gazillion. A Brazilian. A googol (that's a 1 with 100 zeros). A googolplex (that's a 1 with a googol number of zeros). A googolplex plus 1!

Perhaps humankind can form a relay race. As one person gives up, the next person could take over where the last left off. But it turns out that even this strategy is doomed. The universe itself is going to run out of time. (Time, we believe, began ticking at the Big Bang. Once it had started ticking it was believed that it would go on for ever. But recent discoveries about the way the universe is expanding imply that at some point it will be stretched out so much that there will be nothing left to keep the count to infinity going. In fact, there will be nothing left even to keep track of time itself. Time will run out. Time has an end. It too is finite. But that's another story.)

Nevertheless, mathematicians have found some cunning new ways to navigate infinity without having to count all the way to the end. Using ingenious stratagems cooked up at the end of the 19th century, they've discovered not only how to count to infinity but that there are different sorts of infinity. Some bigger than others. It is one of the most extraordinary feats of human endeavour. Getting to the top of Everest only requires a finite number of steps. But, mathematicians have shown how, using the finite equipment in your head, you can hit dizzying heights that make the finite height of Everest pale into insignificance.

So here I am, your mathematical Sherpa, to guide you in our quest to count to infinity and beyond.

You might ask why we want to get there even if we can. You're only ever going to need a finite number of numbers in your day-to-day life. So why worry about getting to infinity? During your lifetime, there will be a biggest number that you will think of and after that you will never think of a bigger number because your finite life will stop you getting any further.

But that is precisely why contemplating the infinite is worth the effort. The infinite provides an escape from the miserable finiteness of our mortal existence. To conceive of the infinite gives those who achieve this feat a sense of transcendence. As the famous German mathematician, David Hilbert, said of the 19th-century mathematician Georg Cantor, who first gave us a view of the infinite, 'No one will drive us from the paradise which Cantor created for us.' It is into that paradise that I wish to take you.

Like a Buddhist monk achieving a state of otherworldliness through meditation, the journey to infinity is going to require tapping into a mathematical Zen-like state of acceptance. There will be moments when that might feel unsettling, but remember that we are trying to access something which may not have a physical reality. The portal to infinity is to be found deep inside the neurons of your mind. But the finite amount of grey matter inside your head is all you will need to reach this infinite mathematical nirvana.

CHAPTER I

How Long Have We Been Counting?

I could sit and count all day,
Sometimes I get carried away . . .

'The Count's Counting Song',
Sesame Street

We started our count to infinity many thousands of years ago. Indeed, it seems that the very first thoughts made by conscious humans probably involved counting. Humans needed to count to keep track of time. The oldest-known evidence of counting is a bone discovered in the early 1970s during an excavation of Border Cave in the Lebombo Mountains between South Africa and Swaziland. The bone is a small piece of the fibula of a baboon and is marked with 29 clearly defined notches. This piece of bone is dated to approximately 35,000 BC. It is believed that bones like these were marking the passage of the days between one full moon and the next.

Another more sophisticated example of counting on bones was discovered between Congo and Uganda and is now housed in the Royal Belgium Institute of Natural Sciences. Called the Ishango bone, it is dated to 20,000 BC and consists of 4 columns of notches which are clearly counting *something*. In one column, we see 11 notches, followed by 13, 17 and then 19. Is it just a coincidence that these numbers are all the prime numbers between 10 and 20 (which is mind-blowingly exciting in itself), or were humans already obsessed with these indivisible numbers? What is clear is that these cave-dwellers were using the notches on bones in order to count.

The cave paintings in Lascaux, which are dated to 15,000 BC, also show evidence of early counting. In addition to the extraordinary images of running animals painted on the cave walls, there are also strange sequences of dots. One hypothesis is that the dots mark the moon's quarter phases. On one such picture, you can find 13 dots which come to an end next to a great picture of a rutting stag. If each dot represents a quarter phase in one lunar cycle, then 13 dots make up a quarter of a year. One season. The picture is probably a training manual for would-be hunters. It is telling them at what point in the year the stags are rutting and therefore are easier to hunt.

The trouble is that a load of dots and notches isn't a great way to count. It's difficult to actually see how many notches you've got on your bone or what number of dots there are on the wall just by looking at them. When you get above 5 dots, humans are not very good at distinguishing just how many dots they are looking at.

In order to be able to take a logical step further, different cultures across the world began to develop more sophisticated ways of counting.

The Ancient Egyptians came up with a string of funny symbols to represent 10 or 100 or higher powers of 10.

They would draw a picture of a heel bone to symbolize the number 10, or one of a frog to represent that they'd got as far as 100,000. But this system wouldn't work very well if you wanted to count all the way to infinity. You'd need new symbols as you got higher and higher. Instead, other ancient cultures discovered the power of what we call the place-value system, one in which you could attempt to count to infinity with only a finite number of symbols.

Evidence of one of these ancient cultures can be found in South America. Around 2,000 years ago, the Mayans started using the system of dots that the cave-dwellers had used, but when the Mayans had written down 4 dots, instead of adding a 5th dot, they did what the prisoner counting days till his release does: they put a line through the 4 dots to indicate 5. When they got to 20, instead of adding more and more dots and lines, they used the place-value system. A second position was introduced to keep track of how many lots of 20 had been counted. See the example below:

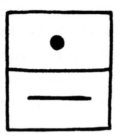

The above diagram indicates one lot of 20 and 5 units. The content of the lower box gives the number of single units (in this case, 5, represented by the line) and the upper box the number of 20s (in this case, one, represented by the single dot). So, this was their way of writing 25. Now that they had created this system, they could count to infinity solely by using dots and lines. They would simply keep on introducing new positions to indicate further powers of 20 (in just the same way that we keep track of powers of 10). The Mayan system of counting, consisting of dots and lines, was very sophisticated. It allowed those Mayans who were interested in astronomy to keep track of vast swathes of time.

This place-value system had already been created and

used by one of the very first cultures to undertake mathematical tasks. The ancient Babylonians counted in 60s. They created symbols that represented all the numbers up to 59, and then they started a new column to indicate an additional lot of 60. At this point, I'm sure you're wondering why we count in groups of 10 while other cultures count in 20s and 60s. The choice of 10 in our decimal system had nothing to do with the special mathematical importance of the number 10, but was solely related to the fact that we counted on the fingers of our hands. The Simpsons, who have 8 fingers, probably count in groups of 8. Maybe the Mayans counted in groups of 20 because they used both the fingers on their hands and the toes on their feet.

So why did the Babylonians count in groups of 60? One reason is because of this number's special mathematical qualities. It is a highly divisible number. You can divide it by 2, 3, 4, 5, 6, 10, 12, 15 and 30. This makes it a very flexible number system. But there is another theory that 60 relates to our human anatomy. There is a way to use the bones in your fingers to count all the way up to 60.

Starting with the thumb on your right hand, this is the digit that you use to do the actual counting. There are 12 bones on the 4 fingers of your right hand. Imagine that you are using your right-hand thumb to number them, pressing your thumb against each bone, all the way from 1 until you reach 12. Next, use all the digits on your left hand (including your thumb), to keep track of how many groups of 12 you have counted. On your left-hand fingers and thumb you can count up to 5 lots of 12 . . . which make a total of 60.

Although we count in 10s today, there are remnants of the Babylonian system of counting in 60s revealed in the way we keep track of time: there are 60 seconds in each of our minutes, and 60 minutes in one hour.

Now that we've got different symbols to keep track of our count to infinity we'll have to come up with a strategy to get there.

CHAPTER 2

Counting Faster and Faster

'An infinity of passion can be contained in one minute, like a crowd in a small space'

Gustave Flaubert, *Madame Bovary*

To warm us up on our way to infinity, here is a possible strategy you could apply for getting there. It's going to involve a bit of maths, but you can't expect to read a book called *How to Count to Infinity* without a bit of an algebraic workout. So, let's limber up and get those mathematical neurons firing.

Here is our strategy. What if I took 8 seconds to count from 1 to 10? Then I'll decide to speed up. I'll count the

13

next 10 numbers from 11 to 20 in 4 seconds. Then the next ten in 2 seconds. Each time I tackle another 10 numbers I halve the time it took me to say the last 10. How long would it take me to get through all the numbers? (Infinitely many of them.)

I've got to add up the times it takes to do each batch of 10 numbers.

$$8 + 4 + 2 + 1 + \tfrac{1}{2} + \tfrac{1}{4} + \tfrac{1}{8} + \ldots$$

That's an infinite number of numbers I've got to add up! Won't that take an infinite amount of time to calculate? And isn't adding up an infinite number of numbers going to give an infinite answer?

Here's our first sneaky strategy to avoid having to do an infinite amount of calculating. Suppose the answer to this infinite sum is 'N'. N might be infinity or it might be some other number, but, whatever it is, we are going to give the answer the name of N.

Now here's one of those magic-trick moments in mathematics that looks like a stupid move at first sight, but when the trick is finished you will realize it was, in fact, an inspired idea.

We are going to calculate what $2 \times N$ is. Multiplying each of the numbers in our infinite sum by 2 we get:

$$2 \times N = 2 \times (8 + 4 + 2 + 1 + \tfrac{1}{2} + \tfrac{1}{4} + \tfrac{1}{8} + \ldots)$$
$$= 2 \times 8 + 2 \times 4 + 2 \times 2 + 2 \times 1 + 2 \times \tfrac{1}{2} + 2 \times \tfrac{1}{4} + 2 \times \tfrac{1}{8} + \ldots$$
$$= 16 + 8 + 4 + 2 + 1 + \tfrac{1}{2} + \tfrac{1}{4} + \ldots$$

Having doubled N, if I now subtract N, I will be left with N once again:

$$2 \times N - N = N$$

But something rather amazing happens when we look at what $2 \times N$ is and then subtract N from it:

$$2 \times N - N =$$
$$16 + 8 + 4 + 2 + 1 + \tfrac{1}{2} + \tfrac{1}{4} + \ldots$$
$$- 8 - 4 - 2 - 1 - \tfrac{1}{2} - \tfrac{1}{4} - \tfrac{1}{8} + \ldots$$
$$= 16$$

Why is the answer 16? Because all the other bits in the first line got taken away by the things in the second line. Only 16 was left. So, we've worked out what N is. N = 16. Therefore, using this strategy, it would only take 16 seconds to count to infinity!

Of course, this strategy requires your mouth to move faster than the speed of light at some point along the way and, unfortunately, the speed of light is finite. We're not going to get to infinity this way. We're going to need to change our line of attack instead.

So, here's our next strategy. We're going to take a leaf out of the way certain indigenous tribes across the globe navigate large numbers, even when they don't have names for these numbers. In many Australian Aboriginal languages there are no names for numbers beyond 5. For example, the Angkamuthi from Cape York count in this way: *ipima* (1), *udhima* (2), and *wuchama* (3). Anything more than 3 is referred to as *makyan* (many).

For the Angkamuthi, the word *makyan* is like their name for infinity, something that they can't count. But even though they can't count beyond 3, they do, nevertheless, have a strategy which helps them work out when one 'infinity' is bigger than 'another'.

Suppose that there are 2 piles of fruit. One pile consists of plums, the other limes. Even though there are *makyan* or 'lots' of fruit in each pile, the Angkamuthi tribesman has a strategy for telling which pile has the most fruit. He starts by taking a plum and a lime from each pile and places them side by side. Now he takes another plum and lime and pairs these two. He keeps on doing this until one pile runs out. If there are fruit left in the other pile he knows that particular *makyan* was bigger than the other *makyan*. If both piles run out at the same time, then it means these *makyan* must have the same size.

This is the same strategy that animals are believed to use. It is very important for animals' evolutionary survival that they have some facility with numbers. Counting allows a bird to know when an egg has been taken from the nest and to be extra-vigilant to ensure that no more disappear. If your troop of monkeys is threatened by a rival troop, then it is important to be able to assess which troop is bigger. The answer will inform your decision to fight or fly. Even though animals are unlikely to have names for numbers beyond 3 (if they have names at all), they apply the same principle as the Angkamuthi; they will mentally pair up monkeys from each tribe until one troop runs out.

The extraordinary insight that the German mathematician Georg Cantor had was that mathematicians trying to contemplate the infinite are not so different from the Angkamuthi or a troop of monkeys. We have names for all the finite numbers. It's when things are infinite that we hit problems. We only have one word – infinity – to describe the size of anything which isn't finite.

But as Cantor realized, that doesn't mean that there might not be more than one infinity, and, if so, that we might be able to compare one infinity with another and judge that they have the same size, or even be able to assess when one infinity is bigger than another infinity. The secret is to find a way to pair up the infinite things in one pile with the infinite things in the other pile and to see if they match up perfectly or not.

So, hold on to your mathematical hats, because we are about to embark on our attempt to count to infinity.

CHAPTER 3

Welcome to the Infinity Hotel

'Meditation is the dissolution of thoughts in
Eternal awareness or Pure consciousness without objectification, knowing without thinking,
merging finitude in infinity'

Swami Sivananda, Hindu spiritual leader

David Hilbert was one of the mathematicians who very
quickly understood the amazing new insight that Cantor
had produced. And in order to try to explain the idea to
others, he came up with a beautiful scenario to help us to
understand the way that Cantor was proposing to compare
infinities. He imagined a hotel with infinitely many rooms.

Each room has a door number, starting at room number 1, followed by room 2, room 3 and so on, carrying on down the corridor to infinity. At the front desk is Concierge Cantor, whose job it is to accommodate all the guests as they arrive at the Infinity Hotel.

Now, imagine tour buses arriving at the Infinity Hotel. Each tour bus will contain an infinite number of guests and each guest will be wearing a badge with their own individual number printed on the badge. The challenge is to understand whether the hotel can accommodate the infinite number of guests that arrive.

For example, the first tour bus to arrive has guests whose badges are numbered with the numbers 1, 2, 3 . . . all the way to infinity. This tour group is easily accommodated because each guest gets paired up with the room number that they are sporting. Nothing too surprising there, and the unfussy guests all leave happily after a few days of chilling out.

Next arrives the Even Numbers tour bus. This tour is for those people who only like even numbers, so each of the guests sports an even number on their badge. So, there is a guest number 2, guest number 4, guest number 6 . . . all the way to infinity. Now at first sight, it looks as if there are half as many guests as there were on the previous tour, so the hotel will be half-empty. But that's not necessarily true. And this is the first subtlety about infinity. Concierge Cantor can actually assign the rooms to the guests in such a way that every room is paired up with a guest. The rule he follows is that the guest takes the room whose number is half the number on their badge. So, guest number 2 takes room number 1, guest 4 takes room 2, guest 6 takes room 3 . . . and so on, all the way to infinity. The hotel is then full, even though it had looked as if half as many guests turned up on this Even Numbers tour bus. Which is good for business, as anyone in the hotel industry would agree.

The infinity of even numbers is the same size infinity as the infinity of all numbers. Think of the Angkamuthi tribesman who pairs up the plums and the limes. If he finds that every plum can be paired with a lime and vice versa, then he can infer that the number of plums and limes is the same, even though he doesn't have a name to count those totals. In the same way, we can say that the infinity of even numbers is the same size as the infinity of all numbers because we've found a way to pair them up perfectly. All the Even Numbers guests are accommodated and the hotel is full, because every room has a guest in it. For example, room number 27 has guest number 2 × 27 = 54 in it.

So, although on paper (or the booking form), it looked like half as many guests turned up on the Even Numbers tour bus, we actually recognize that these two infinities must have the same size. But now here comes a massive tour bus. It holds the Fractions tour group. Every guest has a badge with a fraction on it, and somewhere in the huge crowd, you'll find every fraction on someone's badge. At first sight it looks impossible to house this many guests in the Infinity Hotel. After all, there are infinitely many fractions just between the numbers 1 and 2, and then another infinitely many fractions between 2 and 3. Surely there is no way to accommodate the Fractions tour group in the Infinity Hotel?

But Concierge Cantor is not daunted by the arrival of this group. Miraculously, there is a consistent and logical way to pair up each guest with a room in the Infinity Hotel so that no guest is left without a room. I am going

24

to describe the algorithm, or method, that Cantor came up with for assigning the rooms. It is a subtle idea, but it requires nothing more than a cool head to navigate it.

First of all, Cantor has got to find some way to put some order into this unruly rabble of Fractions guests, all of whom are pushing their way to the front of the queue to get a room. He's got to find some way to make them queue up in an orderly fashion so he can hand out the room keys.

To start with, he pulls out all those Fractions guests who have a 1 on the bottom part of their fraction: ⅟₁, ⅔₁, ⅗₁, ⅘₁ . . . There are infinitely many of them, but he gets them to line up in an infinite queue with ⅟₁ at the start, then ⅔₁, then ⅗₁, and so on, all the way to infinity. But the trouble is that he's still got infinitely many fractions left to deal with, so he can't just hand out the keys to this queue alone.

So, then he makes a second line which runs alongside the first line. These guests comprise all of those with a 2 at the bottom of their fraction. First in line is ½, then ⅖, followed by ⅜ and so on, all the way to infinity. Still there are infinitely many fractions left. But Cantor just keeps on doing this, creating infinitely many more queues like this.

Thus, for example, the 27th line will contain all those fractions with 27 on the bottom, starting with $\frac{1}{27}$, then $\frac{2}{27}$, followed by $\frac{3}{27}$, etcetera.

Every single fraction is somewhere in this line-up. For example: where is guest number $\frac{71}{101}$? She'll be the 71st person in the 101th queue that Cantor lined up. But now, hang on, how does Cantor hand out the keys to the rooms? How does Cantor turn this infinite number of infinite queues into a single infinite queue? Here comes Cantor's stroke of utter genius. Rather than heading down one of the lines, he walks through the infinite grid, snaking out a zig-zag pattern amongst the fractions. He weaves back and forth, making sure that he visits each fraction at some point in his diagonal dance.

For example, fraction $\frac{1}{1}$ is first in the queue and gets room 1. Next Cantor goes to fraction $\frac{2}{1}$, who must receive the key to room 2. But now instead of working his way further down the line, he snakes back to give room number 3 to fraction $\frac{1}{2}$. Let's take a fraction like $\frac{2}{3}$. Which room does this guest get? Well, he will be the 9th guest visited on Cantor's journey, so he gets room number 9.

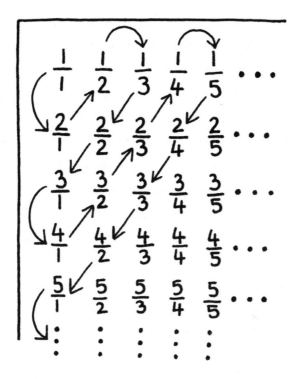

Amazingly, the infinity of guests with fractions on their badges has the same size as the infinity of whole numbers, even though, at first sight, it seemed as if the Fractions guests would totally swamp the Infinity Hotel. But, instead, Concierge Cantor's clever, diagonal, weaving pattern through the guests ensures that every Fractions guest gets paired up with a room.

It is beginning to look awfully like all infinities have the same size. Perhaps every infinite tour bus that turns up can be accommodated? But this turns out not to be the case. Indeed, the Irrational Numbers tour group that was next to appear completely foxed the concierge. However hard he tried, he couldn't find a way to accommodate them all, despite having infinitely many rooms at his disposal. So, just who are these foolish, annoying, irrational guests?

CHAPTER 4

Irrational Numbers

'Number is the ruler of forms and ideas and the cause of gods and demons'

Pythagoras

We have been counting ever since we've been able to think: keeping track of time, deciding whether to fight or fly, rearing our offspring and making sure that none have been stolen from the nest. But more sophisticated mathematics really only grew out of the new demands to measure the world around us which our developing civilizations placed upon us. As new city states emerged around the rivers of the Euphrates and the Nile, the ancient cultures of Babylon and Egypt needed new mathematical tools to shape their environment . . . and to tax it!

Citizens of Ancient Egypt were taxed upon the basis of the size of the areas of land they owned. For the farmer with a rectangular field, that was easy to calculate. The area of his field could easily be worked out by multiplying its length by its width. But the tax authorities had more trouble with those farmers whose fields were beside the Nile. The trouble was that a river doesn't follow straight lines. It meanders. And, as often as not, the land it carves out is in the shape of a perfect half circle. In this way, the tax authorities were faced with trying to calculate the area of a circle and dividing that by two if they were going to be able to tax the farmer correctly.

If the circle had a particular radius, let's call it 'R', then what was the area of the whole circle? The square that surrounded the circle was made up of 4 small squares of the size of R squared, that is to say, $R \times R$ (or R^2). But the area of the circle was smaller than the area of the 4 small squares which made up one large square around the circle, as in the diagram below. The mathematicians of the day discovered that, regardless of how big the circle was, to calculate its area you had to multiply R^2 by the same fixed number each time. This number is what many regard as one of the most important and enigmatic numbers in mathematics: pi.

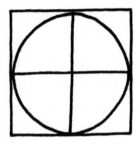

We can find attempts at working out what this number was in one of the first and most significant documents in the history of mathematics: the Rhind Papyrus, which was written by an Egyptian scribe, one Ahmes, in about 1650 BC. Housed in the British Museum, it is full of fantastic mathematics, including the first estimate of a value for pi. In it, Ahmes tries to estimate the area of a circular field, the diameter of which is 9 units across. Because the area of a circle is pi multiplied by the radius of the circle squared, if we know the area and we know the radius, then we can calculate pi.

The Rhind papyrus states that a circular field with a diameter of 9 units is very nearly equal in area to a square with sides of 8 units – but how could this relationship have been discovered? My favourite theory is that the discovery of the value of pi had its origins in the ancient Egyptians' addiction to the game of Mancala. Mancala boards were very popular during this period and were even found carved on the roofs of temples. The boards consist of 2 rows of circular holes.

Each player starts with an equal number of stones, and the object of the game is to move round the board, capturing your opponent's counters on the way. Perhaps Ahmes had been playing Mancala, and as he sat around waiting to make his next move, he might have noticed that sometimes the stones filled one of the circular holes in the Mancala board in a rather nice, decorative way. For example, place 7 stones in any of the circular holes and then you have 1 stone in a central hole surrounded by 6 stones to create a hexagonal shape around the centre stone.

Ahmes might have gone on to experiment with making larger circles and discovered that, if he took the right number of stones to make up an 8 × 8 square, he could then rearrange them into a large circle with a diameter of 9 stones across.

But Ahmes could use this to estimate the area of the circle of which the diameter is 9 units. By again rearranging the stones which had been appropriated by the circle back into the 8 × 8 square configuration he started with, he could estimate the area of the circle by knowing the area of the square (which was 64 units). Remember that the area of a circle is pi multiplied by the radius squared:

$$A = pi \times R^2$$

Therefore, by rearranging the equation we get the following formula for pi:

$$pi = A \div R^2$$

The radius is half the length of the diameter, therefore
4.5 units. So by dividing 64 (the area of the circle) by
4.5 × 4.5 = 20.25 (the radius of the circle squared) we get
that pi is approximately 3.16. Not bad for a first estimate.
As mathematics developed, more cultures tried to capture
this important value, pi, with even greater precision.

Like our Mancala player, the Ancient Greek mathematician, Archimedes, also tried to capture the area of a circle
by using other shapes which he was able to analyse more
easily. By getting a good estimate for the area he could use
this to calculate pi. Archimedes started by drawing a triangle inside and outside the circle. The triangle doesn't look
much like the circle to him. But what if he now doubles
the numbers of sides of the triangles and replaces them with
hexagons instead? Hexagons are a bit closer to circles in
their shape and can therefore be used to get a better estimate for the area of the circle. By continually doubling the
number of sides of the shape which Archimedes was using
to approximate the circle, the difference in area between the
original circle and the many-sided shape he was developing
became smaller and smaller. In fact, we sometimes say that a

circle is a regular polygon with an infinite number of sides. Archimedes didn't go as far as infinity. He stopped when he got to a shape with 96 sides. But, by using this shape he was able to estimate that pi lay between $^{223}/_{71}$ and $^{22}/_{7}$. And it is from Archimedes that we get the approximation of pi that most engineers use, which is $^{22}/_{7}$.

But, as hard as they tried, no one could completely capture this enigmatic number exactly by using a fraction. Any fraction seemed to be either a little too big or a bit too small. The reason the mathematicians of the ancient world were having such a problem is that pi can't be written as a fraction. It isn't a ratio of two whole numbers. It is what we call an irrational number. Although pi wasn't proved to be irrational until the 19th century, these strange new numbers had already been discovered by the Pythagoreans when they were trying to measure another quantity. Not the area of a circle, this time, but the length of a side of a triangle.

Pythagoras was deeply shocked by the discovery that ratios of whole numbers were not sufficient to get the measure of the world. It went against his belief in the music of the spheres, the idea that the universe was made out of perfect whole-number ratios. His belief had arisen from his discovery of the mathematical basis of musical harmony, that the frequencies we find harmonic are all created in a perfect whole-number ratio. But the discovery of these inharmonic or irrational proportions destroyed this Pythagorean philosophy. And the source of this disruption? Pythagoras' own theorem about right-angled triangles.

Take a right-angled triangle whose short sides both have a length of 1 metre: Pythagoras' theorem tells you how to work out the length of the longest side of that triangle, which is called the hypotenuse. Pythagoras says that if you square the lengths of the 2 short sides of the triangle and add these squares together, it is equal to the square of the longest side. The short sides each have a length of 1 metre. Therefore, Pythagoras says that the square of the longer side must be $1^2 + 1^2 = 2$. So how long is the hypotenuse? It's a certain length, which we will call 'L', which, when you square it, gives you the answer 2 (so L is the number which we call the square root of 2). But what *is* this number?

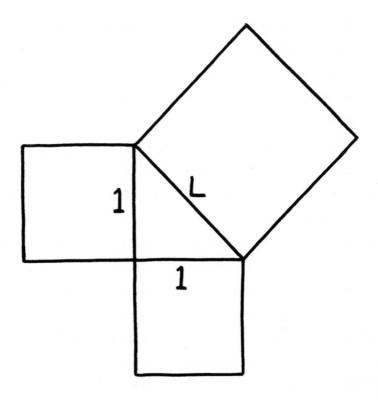

Already the Ancient Babylonians had had a go at trying to work out a value for this number. A tablet housed at the University of Yale dating back to the Old Babylonian Empire that existed from 1900 BC has an estimate for the distance. The tablet is probably the homework of a trainee scribe whose teacher set the task of finding a number whose square was 2. The student didn't do too badly. Written using the sexigesimal system or base 60 used by the Babylonians, they got the length to be:

$$1 + \frac{24}{60} + \frac{51}{60}^2 + \frac{10}{60}^3 = \frac{30547}{21600}$$

which in decimal notation comes out at 1.41421296296 recurring (i.e. in which the last 3 digits, 296, carry on, to repeat themselves infinitely often). It's a pretty impressive estimate and is correct to 6 decimal places. But when you square that decimal number (or that same number expressed as a fraction) the result always just misses being 2. It was one of Pythagoras' followers, Hippasus, who proved that however hard anyone tried to establish the exact number (including that Babylonian scribe!) he (or she) would always find that his or her fraction couldn't be squared to reach 2, because there is no fraction whose square is exactly 2. Hippasus showed that if there was such a fraction, it would imply that even numbers were odd and vice versa. Clearly an absurdity. The only way to resolve this contradiction was to admit there wasn't a number whose square was 2. This new sort of number was called an irrational number, because it can't be written down as the ratio of two whole numbers.

His fellow Pythagoreans were outraged at the discovery. A sect had built up around Pythagoras' belief that mathematical harmony ran the universe. The discovery of these irrational numbers went against their whole ethos. The sect decided to keep the discovery a secret. But Hippasus couldn't resist sharing his great revelation and the news of these irrational numbers began to spread. Having broken the vow of silence imposed on him, Hippasus was taken out to sea and drowned for revealing such disharmony at the heart of the physical world. Who realized that maths could be so dangerous? But these new irrational numbers could not be so easily silenced.

We now understand that numbers like pi and the square root of 2 are examples of numbers which, if you try to write them down as a decimal number, then the decimal expansion continues to infinity, never repeating itself. For example, the square root of 2 written as an infinite decimal starts as:

$$1.414213562\dots$$

and then continues to infinity. If we just take a number with a finite amount of those decimal places then its square will miss 2. It would need all of an infinite number of non-repeating decimal places to truly capture this length on Pythagoras' triangle.

Pi, too, is a number that requires infinitely many non-repeating decimal places to allow it to be captured. Modern mathematical methods and computing power mean that we know pi to a staggering trillion digits. Of course, no one needs to know so many values for the practical purpose of calculating areas. You only need to know pi to 39 decimal places to be able to calculate the area of a circle the size of the observable universe in such a way that the error in the calculation will be of the order of the size of a hydrogen atom. But to truly know pi you need to be able to count to infinity.

It's amazing that just looking at a circle or a triangle you are in fact staring into the infinite.

But pi and the square root of 2 are just a couple of an infinite number of these possible infinite decimal numbers. However, the extraordinary revelation that Cantor made was that *this* infinity – out of all infinite decimal numbers – is a genuinely bigger sort of infinity than all the whole numbers or fractions so beloved of the Pythagoreans.

CHAPTER 5

My Infinity Is Bigger
than Your Infinity

'Whoever you are, go out into the evening,
leaving your room, of which you know every bit;
your house is the last before the infinite,
whoever you are'

Rainer Maria Rilke, 'Initiation'

Let us return to the Infinite Hotel, where the concierge is trying to cope with the latest tour bus that has just rocked up containing the Irrational Numbers tour group. The guests on the bus have all got badges with individual numbers on them, but this time each number is one of those numbers with an infinite decimal expansion like pi or the square root of 2. The tour organizer is absolutely convinced that Cantor's Infinite Hotel should be able to accommodate all his holiday-makers. There are certainly infinitely many guests on the bus, but Cantor's hotel has infinitely many rooms. However, the concierge is starting to get worried. He's realized that, however hard he tries to match up the rooms with the guests, he can prove that there will always be guests *without* a room.

The tour operator is not going to be defeated by Concierge Cantor's pessimism. After all, they had managed to get all the Fractions tour group members into the hotel the night before and that looked like a totally impossible task when they first arrived. But it just depended on finding a clever way to put the Fractions guests into some order so that the concierge could make a weaving route through them to ensure that each got a room.

The tour operator claims he has a really clever algorithm that will do the same for the Irrational Numbers group on the tour bus. He starts lining them up using his algorithm and claims that by the time he's finished he'll have got all the guests to form an orderly line and then the concierge just has to go down the queue handing out the room keys. Let's suppose that the tour operator's algorithm has got pi at the front of the queue. The square root of 2 is next, and after that the roll-call goes on from one infinite decimal number to another:

1 gets paired with 3.1415926 . . .

2 gets paired with 1.4142135 . . .

3 gets paired with 2.7182818 . . .

4 gets paired with 1.6180339 . . .

5 gets paired with 1.2020569 . . .

The tour operator is confident that his algorithm has ensured that every infinite decimal is going to be somewhere in the queue. But Concierge Cantor is not convinced. He starts to cook up an infinite decimal number that he can prove can't be anywhere in the queue.

He begins with the first person in the queue. 'What's your first decimal place?' he asks. '1' the first guest replies. Well, the concierge then writes down a '2' as the first decimal place of the number he is cooking up. That will make sure his decimal number isn't the same as that of the first person in the queue because their numbers differ in the first decimal place. Next, he moves on to the second person in the queue. Concierge Cantor is going to use this person to decide what to put down as the second decimal of the number he is cooking up. 'What's the second decimal place in your number?' he asks the second guest. The second guest looks at his badge and says '4'. The concierge therefore chooses 5 for the second decimal place of his new number. That way he can ensure that his number is different from the infinite decimal on the second guest's badge. Why? Because the second decimal place of both the guest and the concierge's numbers are different.

Hopefully, you've guessed what Concierge Cantor is going to do. He works his way down the queue and uses the *nth* guest to choose the *nth* decimal place of the number he is cooking up. The trick he uses is to choose a different number to go in the *nth* decimal place of his number as compared with the *nth* decimal place of the number belonging to the guest.

Once he's done this, he comes back to the tour operator and says: 'You've missed the guest with *this* number on his badge. He's nowhere in the queue.'

The tour operator isn't going to give in too quickly. 'Are you sure? Try the 71st person in the queue.'

They head down to the guest, and the concierge tells the tour operator to check the 71st decimal place of this guest's number. When they compare the answer with the 71st decimal place of the concierge's missing guest, sure enough, the numbers in the 71st decimal places of both are different. That is, after all, how the concierge cooked up his number. It doesn't matter how hard the tour operator tries, the way the concierge has built the infinite decimal number guarantees that the guest with this number on their badge is nowhere in the queue.

The tour operator finally admits that this guest is missing from the queue. 'OK! OK! So, let's just put him at the front of the queue and shunt all the others along.' But the concierge isn't happy. 'Look, I can just play the same game again and cook up another missing guest. It doesn't matter how hard you try, my trick shows your queue is always missing a guest. Not just one guest, but it's missing infinitely many guests.'

And that's the crux of the problem. It doesn't matter how clever the tour operator's algorithm is, no matter how hard he tries to find some order in the infinite decimals, Concierge Cantor's counter-strategy shows that a queue of infinite decimal numbers will always miss out some of those guests. The infinity of infinite decimals cannot be paired up with the infinity of whole numbers. It is a genuinely bigger sort of infinity. This infinity is called uncountable, because it can't be paired with the numbers which *can* be counted.

This, to my mind, is one of the most staggering moments in mathematical history. Infinity had previously been regarded as something beyond knowledge, something that couldn't be understood. All infinities were lumped together in the mathematician's 'lots'. But Cantor had found a way to navigate infinity and to show that infinity wasn't just one thing. That there were many sorts of infinity. Some bigger than others. We could give these infinities names. 'Lots' suddenly became numbers with names in their own right. It was as if we'd seen the world in black and white,

48

and then Cantor showed us that the mathematics of infinity is actually multi-coloured. It is like the colour white being pulled apart to reveal an infinite spread of different-coloured infinities.

The trick was not to start counting, '1, 2, 3,' and then to hope to reach infinity. Instead, a change of perspective allowed us to think of infinity in one go and, by doing so, to show that infinity is a many-headed beast. Amazingly it took just 48 pages for us to get to infinity. That's the power of mathematical thought. Using the finite equipment in our head we can transcend our finite surroundings and touch the infinite. At the end of our journey to infinity, I hope you feel you can share Hamlet's feelings when he famously declared: 'I could be bounded in a nutshell and consider myself king of infinite space' (Shakespeare, *Hamlet*, Act 2, scene 2).

So, as we leave the transcendental realms of the infinite and return to our mundane, finite existences, are there any implications for our daily lives of this mathematical ability to navigate the infinite? Amazingly, there are. Much of modern life is underpinned by similar tools to those we have been cooking up on our journey through the infinite. To navigate infinity, we have been devising algorithms that allow us to pair up two seemingly different infinite sets and show they have the same size. This ability to come up with a method or recipe to pair things up, no matter how large the number, is at the heart of many of the algorithms that control our daily lives.

Mathematical algorithms are helping us to navigate pathways through a multitude of things: finding our way in our cities, searching for the website that we were looking for or recommending books and films we might like. Mathematics is even pairing us up with people who might make great partners for life. So those tools that gave us a way to match up every fraction with every whole number might one day be just the mathematics that will help to pair you up with the love of your life. Now, that *has* to be worth a trip to infinity and back . . .